INTRODUCTORY COLLEGE MATHEMATICS

ROBERT D. HACKWORTH, Ed.D.
Department of Mathematics
St. Petersburg Junior College at Clearwater
Clearwater, Florida

and

JOSEPH HOWLAND, M.A.T.
Department of Mathematics
St. Petersburg Junior College at Clearwater
Clearwater, Florida

Theodore Lownik Library
Illinois Benedictine College
Lisle, Illinois 60532

SAUNDERS SERIES IN MODULAR MATHEMATICS

Computers

W. B. Saunders Company: West Washington Square
                          Philadelphia, PA   19105

                          12 Dyott Street
                          London, WC1A 1DB

                          833 Oxford Street
                          Toronto, Ontario M8Z 5T9, Canada

510
H123i
v.15.

Photographs courtesy IBM World Trade/Americas Far East Corporation.

INTRODUCTORY COLLEGE MATHEMATICS   ISBN 0-7216-4424-4
Computers

© 1976 by W. B. Saunders Company.   Copyright under the International Copyright Union.   All rights reserved.   This book is protected by copyright.   No part of it may be reproduced, stored in a retrieval system, or transmitted in any form or by any means, electronic, mechanical, photocopying, recording, or otherwise, without written permission from the publisher.   Made in the United States of America.   Press of W. B. Saunders Company.   Library of Congress catalog card number 75-23631.

Last digit is the print number: 9 8 7 6 5 4 3 2 1

PREFACE

Computers

This book is one of the sixteen content modules in the Saunders Series in Modular Mathematics. The modules can be divided into three levels, the first of which requires only a working knowledge of arithmetic. The second level needs some elementary skills of algebra and the third level, knowledge comparable to the first two levels. *Computers* is in level 3. The groupings according to difficulty are shown below.

Level 1	Level 2	Level 3
Tables and Graphs	*Numeration*	*Real Number System*
Consumer Mathematics	*Metric Measure*	*History of Real Numbers*
Algebra 1	*Probability*	*Indirect Measurement*
Sets and Logic	*Statistics*	*Algebra 2*
Geometry	*Geometric Measures*	*Computers*
		Linear Programming

The modules have been class tested in a variety of situations: large and small discussion groups, lecture classes, and in individualized study programs. The emphasis of all modules is upon ideas and concepts.

Computers is appropriate for all non-science students especially education, business, and liberal arts majors. The module is essential for math-science and technical students with a need for understanding some basic knowledge of computers.

The module begins by presenting the early history of computing machines emphasizing the effect of need on their development. Then the emphasis is on switching circuits and computer applications. *Computers* ends by exmplaining the basic components of a computer, the elementary skills of flowcharting, and BASIC programming language.

In preparing each module we have been greatly aided by the valuable suggestions of the following excellent reviewers: William Andrews, Triton College, Ken Goldstein, Miami-Dade Community College, Don Hostetler, Mesa Community College, Karl Klee, Queensboro Community College, Pamela Matthews, Chabot College, Robert Nowlan, Southern Connecticut State College, Ken Seydel, Skyline College, Ara Sullenberger, Tarrant County Junior College, and Ruth Wing, Palm Beach Junior College. We thank them and the staff at W. B. Saunders Company for their support.

Robert D. Hackworth
Joseph W. Howland

NOTE TO THE STUDENT

OBJECTIVES

Upon completion of this module the reader is expected to be able to demonstrate the following skills and concepts:

1. To be able to indicate the basic features of a computer.

2. To be able to state the difference between a calculator and a computer.

3. To be able to state the needs that prompted computer development.

4. To be able to identify the early computer pioneers and their work.

5. To be able to indicate when current will flow through a switching circuit.

6. To be able to categorize computer applications.

7. To be able to draw a flowchart from a simple algorithm.

8. To be able to write a program in BASIC from a simple flowchart.

Three types of problem sets, with answers, are included in this module. Progress Tests appear at the end of each section. These Progress Tests are always short with only four to six problems. The questions asked in Progress Tests always come directly from the material of the section immediately preceding the test.

Exercise Sets appear less frequently in the module. More problems appear in an Exercise Set than in a Progress Test. Section I of the Exercise Sets contains problems specifically chosen to match the objectives of the module. Section II contains challenge problems.

A Self-Test is found at the end of the module. Self-Tests contain problems representative of the entire module.

To promote learning, the student is encouraged to work each Progress Test example and Exercise Set in detail as it is encountered, checking each answer, and reviewing when difficulties are encountered. This procedure is guaranteed to be both efficient and effective.

CONTENTS

Introduction..1

Early History..2

How Switches Implement Computer Decision Making..............15

Switches and Adding..20

Adding Three One-Digit Binary Numerals.......................26

Stored Programs, Transistors, and Other Developments.........30

What is a Computer?..37

Flowcharts and Algorithms....................................41

Flowcharts Expanded..45

Computer Languages (*How to Talk to a Computer)..............56

Flowcharts and BASIC Language................................59

Module Self-Test...68

Progress Test Answers..73

Exercise Set Answers...75

Module Self-Test Answers.....................................82

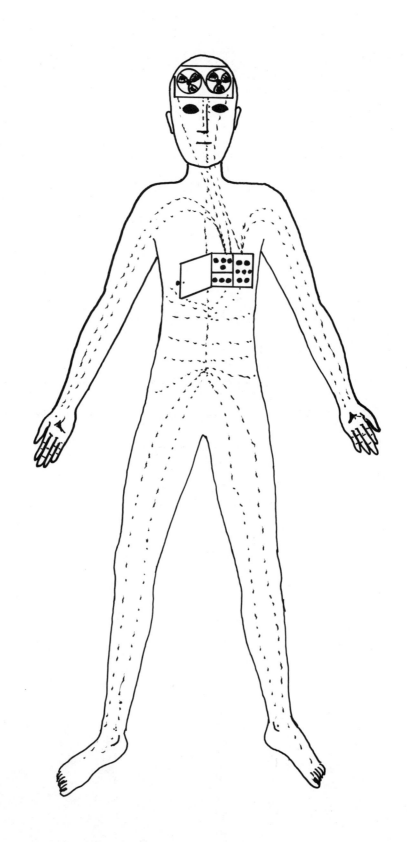

COMPUTERS

INTRODUCTION

"Computers are going to think for us."

"People are going to be replaced by computers."

"The computer goofed."

"Computers will take over the world."

The previous statements are some of the myths existing about computers. When the module is completed, the reader will realize that some of these statements are partially true and others completely false.

A computer is a machine that assists man in solving problems. One of a computer's most important features is its ability to follow a set of instructions until the problem is solved. A typical problem is that of receiving, storing, and printing out on demand the records of traffic violations of drivers for a State Bureau of Public Safety. If done by hand the recording, filing, and recalling of traffic violations is a time consuming job, yet the procedure itself is not difficult. Such a job is ideal for a computer; it never gets tired or bored and can do hundreds of thousands of consecutive operations without a mistake. Computer storage of information is one illustration showing how computers fill a need. The following history of the computer will show that the evolution of computers came about because men sought easier and better ways to solve problems.

EARLY HISTORY

As primitive man learned to count, fingers could be considered part of one of the original computers. A cave man may have lifted a finger each time a woman ran past. In this way, he was putting data or input into his computer (brain). His fingers also served a memory function. When his buddy asked him how many women had run into the glade, he could look at his fingers for the answer. What these cave people did after this early streaking exercise is not in the realm of this module. That is because it concerns the various uses of computer produced data.

When the memory capability of the cave man's brain was exceeded, sticks, stones, notches on sticks and other devices were used to aid in counting. All these devices provided a way to put information into the machine's "input," hold the information in the "memory" and give the answer as "output."

As man's activities became more sophisticated, his counting needs became more complicated. One of the devices that was developed to help solve mathematical problems was the abacus. Its earliest form was probably rows of grooves in a smooth sand table. Stones were placed in the grooves and moved toward one end of the groove as objects were counted. A stone in the first groove could represent a one, in the second groove each stone represented a ten, in the third groove each stone represented one hundred, and so forth. Therefore, the abacus had the capacity to represent large numbers in a compact form. Adding, subtracting, multiplying and dividing can all be done quickly on the abacus. In fact, in its present form consisting of a frame holding rows of beads, a skilled abacus operator can calculate faster than an electric calculator. However, the most skillful abacus operator cannot keep pace with a person using the cheapest of the modern hand-held electronic calculators. There is evidence that the abacus has been used from about 500 B.C. to the present day.

An abacus.

Abaci are widely used hand powered machines to aid man in computation. They have some of the features of the modern electronic computer: input, memory, and output. But they do not have the distinguishing features of the electronic computer: the ability to automatically follow a set of instructions and to make decisions.

Another ancient machine is called the Antikytheria Device. It was probably developed a few hundred years after the abacus. As a calculator, it was probably not widely known because the only model of the device (and possibly its inventor) was lost in a shipwreck off the island of Antikytheria in the Mediterranean Sea. The Antikytheria Device was a primitive form of a planetarium, consisting of brass gears, axles, and sliding rings. It was used to locate the position of the stars. It even had a sliding adjustment to compensate for the quarter of a day gained every four years in the calendar. The setting of the device, when it was found, agreed with the location of the stars in the year 80 B.C. according to present calculations. Other evidence in the wreck also indicated the ship had sunk in 80 B.C. The existence of gears and axles in the device showed the advanced technology of the time. These skills seem to have been lost with the rise of the Roman Empire, as no other computing device was developed for about 1700 years.

4   Introductory College Mathematics

An early calculating machine called Napier's bones or Napier's rods invented by John Napier (1550-1617) could perform multiplication. It consisted of lengths of bone or ivory each inscribed with a number from one to ten. Multiples of the number were arranged in diagonal fashion below the number as shown in the figure below. By sliding the rods up and down in relation to each other, multiplication answers can be easily found.

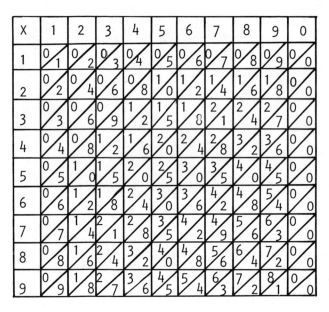

Napier's Bones

To multiply 87 by 4, put the X box by the 8 and 7 boxes. Both 32 and 28 are next to 4 on the X bone. How are they used to find the product? A hint for the procedure may be found in the calculation on the right.

```
 87
 x4
 28
 32
---
348
```

The product of 489 and 9 may be found by arranging bones as shown on the right.

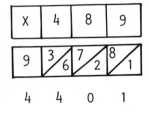

```
 489
 x 9
 ---
  81
  72
  36
----
4401
```

Besides invention of the simple calculator mentioned previously, Napier is credited with much of the development of logarithm mathematics. His work led to the invention of the slide rule; another simple calculator which can find products, quotients, powers and roots using the basic laws of exponents (logarithms). Long a familiar sight at the belt of the engineering students, the slide rule has now been superseded by the hand-held electronic calculator. The slide rule's effectiveness is no match for the speed, accuracy, and economy of the hand-held calculators.

Like the "bones" shown previously, answers are found on the slide rule by moving the center part of the rule so that the appropriate numbers are matched. The "slip stick" below is in position to show that 214 (under 1 on the C scale)• 194 = 41516 (under the hairline).

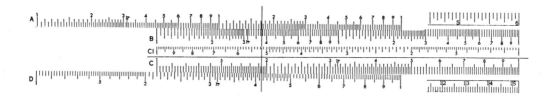

In 1642, Blaise Pascal built a mechanical calculator as an aid for relieving the wearying routines of his tax accountant office. His invention consisted of a series of toothed wheels arranged side by side. Each wheel had ten teeth to correspond with the digits 0 to 9. To enter a 3 into the machine, Pascal moved the first wheel to the third tooth. To add 4 to 3 he moved the wheel 4 more teeth. The invention he made to handle the situation when a sum was 10 or greater is still used today in counting devices such as speedometers. When the sum of two numbers contained two digits the second wheel was advanced one tooth by a projection placed on the first wheel. Each wheel had a projection on its side so that as it moved from 9 to 0 the projection advanced the next wheel one position. Because of this, a tooth on the first wheel represented the number one. A tooth on the second represented ten. A tooth on the third wheel represented one hundred and so forth. Consequently, Pascal's calculator and the abacus could represent large numbers. Like the abacus, the calculator had input, memory and output features as do modern computers.

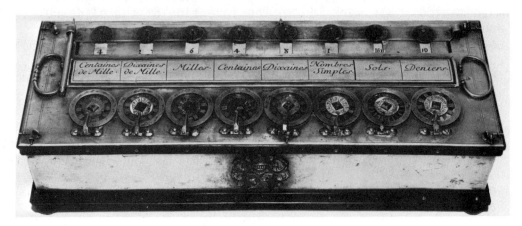

Pascal's calculator.

The next advancement in calculator design was made by Gottfried Wilhelm Leibniz, a German mathematician and philosopher. Leibniz, besides being the creator of the calculus along with Isaac Newton, built a mechanical calculating device using stepped cylinders. The stepped cylinders made multiplication an operation of successive addition. Leibniz's machine was not widely used, but shows the level of invention and technology relative to computational devices of his time.

Leibniz's calculator.

In 1820, a German, Hahn, built a practical variation of Leibniz's machine that could add, subtract, multiply and divide. Copies of this machine were used all over Europe and the United States. The desk calculators used today, except for the electronic calculators, are direct descendants of Hahn's machine.

The next step in the evolution of computers was prompted by the sad condition of the numerical tables used in the United Kingdom. Numerical tables were needed in architecture, surveying, navigation, and accounting. Since the calculators in every table were done by hand they contained many errors. Some of the errors were discovered, to their dismay, by ship captains when their ships wrecked because of incorrect figures in the navigation tables. As someone remarked, it seemed a desperately extravagant way to find errors in the tables. Luckily for the other seamen, an Englishman, Charles Babbage, in 1792, became interested in producing error free tables. He found that the original tables had been constructed by using three groups of people as a team to calculate the entries. The first group chose the proper procedure and formulas, the second group organized the computation in the correct sequence of operations, and the third actually proceeded with the computation.

Babbage wanted to compute the tables entirely by machine thereby eliminating errors from the tables. He found that all the entries could be figured using only the operation addition. He called his proposed machine a difference engine because it would produce answers by adding the correct differences between entries. The table below shows how successive differences can be used to build a table of squares.

NUMBER	SQUARES	DIFFERENCES	DIFFERENCES
1	1		
2	4	3	2
3	9	5	2
4	16	7	2
5	25	9	2
6	36	11	2
		13	2

8 Introductory College Mathematics

Notice that every entry in the column of squares can be reached by adding the differences in proper sequence. Working the tables backwards from Difference to Squares was one of Babbage's contributions to table building.

Babbage's difference engine.

Progress Test 1

1. Make a table of cubes similar to Babbage's table of squares so that the table could be extended forever by adding the differences shown in the table. (Hint: Make three difference columns.)
2. Write a short research paper on Charles Babbage, Gottfried Leibniz, Blaise Pascal and John Napier.

Even though Babbage obtained money from the government to build his machine and made very precise drawings of every part required, the machine was never finished. Before his first machine could be built, Babbage had already moved on to design what he called an analytical engine that would be able to solve many different types of calculations.

The analytical engine was a general purpose computer rather than a special purpose machine like the difference engine. Strangely enough, the development of an automatic loom in France by Jacquard, patented in 1801, played an important part in Babbage's plans for his engine. Jacquard's revolutionary invention was to use punched cards to control the loom as it wove a design into cloth. The arrangement of the holes controlled the action of the loom to create the design desired. Jacquard, by using the punched cards to control the bobbin, was effectively programming the loom's actions. Punched cards have played a major part in computer operation since that time. The inventive Babbage saw the possibilities of using punched cards to control his analytical engine. It was said that Babbage wove mathematics with his machine just as Jacquard wove cloth with his loom.

Babbage called the calculating part of his machine "the mill," and the part that held the data "the store." Today, the mill would be called the arithmetic unit and the store would be called the memory. He planned to feed instructions and data into the mill using punched cards, a feature not possessed by previous calculators. The data would be held in the store to be recalled by the mill as indicated by the instructions. The answers or output of the engine would automatically be recorded on punched cards. Babbage's engine was entirely mechanical. Every operation was done by gears, cams, and axles, giving fifty place accuracy in its answers. Sadly, Babbage's ideas were years ahead of the mechanical technology of his time, so a model of the engine was not built until after his death. The years from 1820 to 1880 saw many advances in machine and electrical technology that made the implementation of Babbage's ideas more feasible.

10   Introductory College Mathematics

Jacquard's loom.

The difficulties with tabulations involved in organizing the data from the United States census was the next problem that prompted further invention of computer design and technology. The tabulation of the 1880 census took seven years to complete manually. James Powers and Herman Hollerith, working for the census bureau, thought that twenty years might be required to tabulate the 1890 census. Consequently, these two men began development of a punch-card tabulating machine. Using cards punched to show certain information, the machine would take the data off the cards electrically by letting wires through the holes to make contact with a pool of mercury. The electricity flowing through the wires to the mercury would register on the dials of the tabulating machine. The practicality of the machine for tabulating data created a demand for its use in many businesses and governmental units. Hollerith formed a company which later became International Business Machines (IBM) and Powers founded a company that became Sperry Rand Corporation.

Hollerith's tabulating machine.

The record keeping necessary to implement the Social Security Act of 1935 created a tremendous need for data handling. In addition, the technical needs of World War II created a fantastic amount of information that had to be handled efficiently. Personnel records, logistics inventory control and problem solving needed for weaponry development are some of the needs that prompted rapid computer development during the late 1930's and 1940's.

In 1937, Dr. Howard H. Aikens, after working several years building a computer, rediscovered Babbage's ideas about a computing machine, and, like Babbage, had the technology to build it. When Aikens read Babbage's material he felt that Babbage was speaking directly to him. Assisted by IBM and using punched cards and Hollerith's tabulating machine, Aikens built the Mark 1 Automatic Sequence Controlled Computer. (A computer is a machine that follows a set of instructions. A calculator is not a computer because it cannot follow a set of instructions.) Aiken's machine weighed five tons and contained seventy-eight adding machines and calculators wired together with 500 miles of wire. Data was fed to the input by switches and punched cards. Output allowed for twenty-three digits and was from electric typewriters or punched cards. There were seventy-two arithmetic registers serving as memory to hold numbers until needed in the calculations. After its completion, Mark 1 was used twenty-four hours a day for about fifteen years. Mark 1 was the first operational automatic computer and therefore dates the beginning of the modern computing age of 1940. Mark 1 was classed as an electro-mechanical computer and contained no vacuum tubes or transistors. When the computer made a decision, the switch was thrown by an electric relay. A relay is an electro-magnet that requires about one-tenth of a second to operate making an audible click. Observers reported that when Mark 1 was operating the clicking of the relays sounded like a room full of ladies knitting. The same decision-making functions of a relay can be accomplished with vacuum tubes or with transistors, which were invented in 1946.

ENIAC (Electronic Numerical Integrator and Computer) designed by Dr. Mauchly at the University of Pennsylvania and Presper Eckert, a student, was put in operation in 1945. ENIAC contained 18,000 tubes, 500,000 soldered joints, and 6,000 switches, and had a memory for storing numbers. Its operations were under the control of instructions that were wired into another part of the machine. Several days were required to wire in the instructions, meaning that the computer was completely out of operation while the hundreds of connections were made. The vacuum tubes of ENIAC could perform operations similar to the relays of Mark 1 but in one-millionth of a second. Although the vacuum tube increased the speed of the computer operation from ten operations per second to 1,000,000 operations per second, they frequently failed to work correctly. The computer rarely ran for more than thirty minutes without "down time" for repairs. Another problem with vacuum tubes is the heat they produce. ENIAC gave off furnace-like heat, shortening the life of all its components. Even though maintenance was a problem, ENIAC was successful and was used primarily to compute trajectories and firing tables for the artillery pieces of the U. S. Army as World War II drew to a close.

Relays, vacuum tubes and transistors all perform a similar function. They can act as switches to either allow current through a wire or cut it off.

Exercise Set 1

I. True or false.

1. Computers are not suited to do repetitious jobs.

2. Viewing a human as a computer, the fingers serve as the calculator and the brain serves as both input and output.

3. On the abacus, every bead represents one dollar, when counting dollar bills.

4. In the ancient abacus pictured on the right, the decimal numeral 124 is represented by the 7 stones.

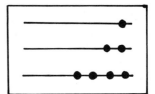

5. The digits on the fourth wheel on Pascal's calculator indicated numbers of thousands.

6. Babbage saw a way to construct mathematical tables using only one arithmetic operation.

7. Jacquard used punched cards to give instructions to his computer.

8. Hollerith and Powers invented the first punched card tabulating machine as a hobby.

9. Babbage had the pleasure of seeing his machine become widely accepted.

10. The vacuum tubes of ENIAC operated about 100,000 times faster than the relays of Mark 1.

11. ENIAC was built to handle the data created by the Social Security Act.

12. Mark 1 was the first operational automatic computer.

13. ENIAC was fed instructions by a deck of punched cards.

14   Introductory College Mathematics

14. Computers can follow a set of instructions automatically and calculators cannot.

15. Fill in the spaces below to show the correct setting of Napier's bones for the multiplication of:

   a. 7 and 569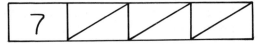

   b. 8 and 2468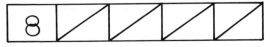

16. How much time would Mark 1 need to complete 1,648 operations?

17. How much time would ENIAC need to complete 1,648,000 operations?

II. Challenge Problems

1. Copy the slide rule pictured below and set it to show the following problems.

   a. $2 \cdot 3 = 6$ b. $1.5 \cdot 1.5 = 2.25$
   c. $37 \cdot 193 = ?$ d. $255 \cdot 47 = ?$
   e. $625 \div 25 = ?$ f. $85 \div 5 = ?$
   g. $0.005 \div 20 = ?$

2. How much time would be saved by using ENIAC to complete 1,562,478,000 operations instead of using Mark 1?

3. How many "carry over" operations would have to be done by Pascal's calculator to complete the following addition?

$$38 + 92 + 284 =$$

HOW SWITCHES IMPLEMENT COMPUTER DECISION MAKING

Switches play an important role in computers. Some switches allow power to flow to the major sections of the computer and are turned on and off by the computer operator when the computer's operation is desired. Many of the switches in a computer are embedded deep within the labyrinth of wiring in the computer. They give the computer the ability to make choices. These are the switches discussed in this section.

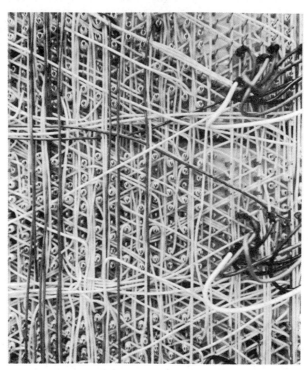

Computer circuitry.

A switch in an electrical wire is analogous to a valve in a water line. Just as a valve allows water to be turned off and on, when a switch is closed the current will flow and when it is open the current cannot flow.

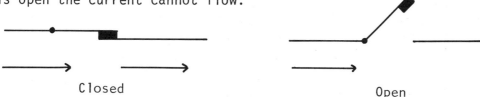

Closed   Open

The current does not flow through the open switch because of the gap in the line.

16   Introductory College Mathematics

A switch's open-closed position is controlled by the instructions in the computer. If a switch controlling the adding section of the computer is closed by the instructions, the current will flow to the adding section and the operation will occur; however, if the switch is kept open by the instructions, the current will not flow and addition will not be done.

The computer's program opens and closes the switches according to the logic of the processes the computer is following. For example, the area of a circle is found by the formula $A = \pi r^2$. If the computer program is for computing the areas of circles, the program will instruct the machine to multiply the radius of the given circle by itself and then multiply the result by pi. The "and" was underlined in the previous sentence because of the importance of the word. The radius is to be multiplied by itself and then multiplied by pi. Two switches will be involved in this operation; one for each multiplication. The switch controlling the multiplication by r will have to be closed and the switch controlling the multiplication by pi will also have to be closed. Which arrangement of the switches below will produce the correct action of the computer?

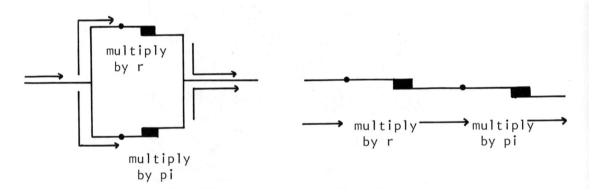

The diagram on the right will cause the machine to multiply by r and then to multiply by pi. But the diagram on the left will multiply by r or multiply by pi or both because there are two ways for the current to travel. Consequently, the diagram on the right is called an "and" gate and the diagram on the left is called an "or" gate. (A gate is an appropriate analog for a switch as they both can be closed or open.) If three switches, p, q, and r control three operations in the computer, then, when the program calls for p to be done and q and then r to be done, the three switches will be arranged in an "and" arrangement (series arrangement) as shown on page 17.

Current flowing in from the left will activate operation p first, and then operation q, and then operation r. In an "and" gate the current does not flow unless the first switch is closed and the second switch is closed also. If either switch is open, the current will not flow.

You may remember from a study of logic that an "and" statement is only true when both parts of the statement are true. In which of the four arrangements below will the current flow completely from left to right?

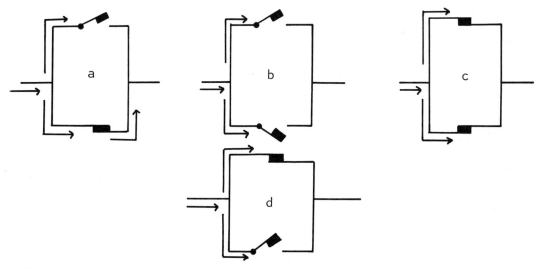

The current will flow through diagrams a, c, and d because in each of those cases either one switch is closed or the other is closed. When one or both switches are closed in the above examples the current will flow through the switches. Such arrangements are called "or" gates and are said to be wired in parallel.

"And" and "or" gates can be arranged in many ways depending on the sequence of operations desired in a computer program (set of instructions). Using p, q, and r to represent three different operations in the computer, the instruction "do p and (q or r)" would be accomplished by the arrangement in the diagram on page 18.

18   Introductory College Mathematics

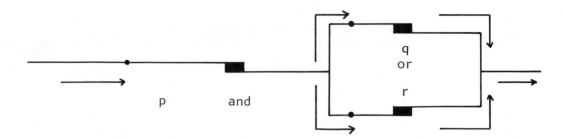

Notice that the current must go through p and then goes through either q or r.

Switches can be arranged in pairs so that when one of a pair is closed its mate is open. In other words, the condition of one switch in a pair is the opposite of the other switch. When this relation is desired between two switches, one can be designated p and its opposite -p. Using this notation, when switch q is closed, then switch -q is open. Using p and q to represent switches that are closed, which examples below will allow the current to flow from one end to the other?

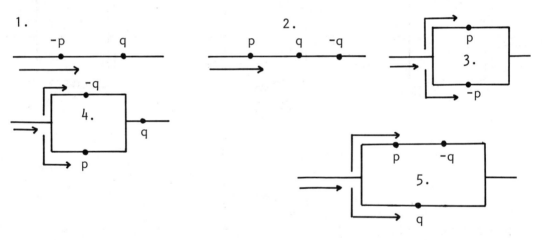

The current will flow through 3 because it can go through switch p in the upper branch of the diagram. The current can flow through the lower branch of 4 through p and then q as both of them are closed.

The current can go only through the lower branch of 5 since -q means an open switch. Neither diagram 1 or 2 will allow the passage of current as they both contain an open switch. Interpreted in terms of "and" and "or," 1 and 2 are "and" gates with one switch open. Therefore, no current can pass. Gates 3, 4, and 5 are "or" gates with one switch closed. Consequently, current will flow through these gates.

Progress Test 2

True or false.

1. The diagram below shows a switch in a closed position.

2. Electric current will flow through an open switch.

3. Electricity will flow through the circuit below.

4. Electricity will flow through the diagram below.

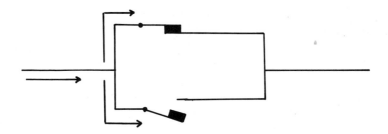

5. The diagram below shows the arrangement for "p and q."

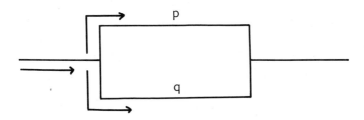

20   Introductory College Mathematics

6. The diagram below shows the arrangement for "a or b."

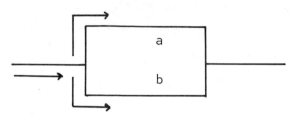

SWITCHES AND ADDING

Adding can be done using the base ten numerals that are familiar, but ten digits 0, 1, 2,...,9 are required. Addition can be done in any base. A base convenient for computer use is base two because there are only two digits necessary in base two; 0, 1. The current at a point in the computer circuit is in only one of two states: either on or off. Consequently, the condition of current "on" can be assigned to the digit "1" and the condition "off" can be assigned to the digit "0."

The numeral 110 in base ten means one one-hundred, one ten, and no ones, and equals one-hundred ten in base 10 (110_{10}).

$$110_{10} = 1 \cdot 10^2 + 1 \cdot 10 + 0 = 110_{10}$$

The numeral 10_2 in base two means one two, and zero one's and equals two in base ten (2_{10}).

$$10_2 = 1 \cdot 2 + 0 = 2_{10}$$

The numeral 11_2 in base two means one two, and one one and equals three in base ten (3_{10}).

$$11_2 = 1 \cdot 2 + 1 = 3_{10}$$

The numeral 1111_2 stands for the number 15_{10} because

$$1111_2 = 1 \cdot 2^3 + 1 \cdot 2^2 + 1 \cdot 2 + 1 = 8 + 4 + 2 + 1 = 15_{10}$$

similarly,

$100_2 = 1 \cdot 2^2 + 0 \cdot 2 + 0 = 4_{10}$

$101_2 = 1 \cdot 2^2 + 0 \cdot 2 + 1 = 5_{10}$

$110_2 = 1 \cdot 2^2 + 1 \cdot 2 + 0 = 6_{10}$

$111_2 = 1 \cdot 2^2 + 1 \cdot 2 + 1 = 7_{10}$

$1000_2 = 1 \cdot 2^3 + 0 \cdot 2^2 + 0 \cdot 2 + 0 = 8_{10}$

$1001_2 = 1 \cdot 2^3 + 0 \cdot 2^2 + 0 \cdot 2 + 1 = 9_{10}$

The results of adding two one-digit binary numerals are listed below.

$$1 + 1 = 10_2 \quad 1 + 0 = 1 \quad 0 + 1 = 1 \quad 0 + 0 = 0$$

The addition table that shows the addition facts above is

+	0	1
0	0	1
1	1	10_2

Progress Test 3

Add the base two numerals below.

1. $1 + 10_2$
2. $1 + 1$
3. $1 + 0$
4. $1 + 1 + 1$
5. $1 + 0 + 1$
6. $101_2 + 11_2$

22 Introductory College Mathematics

Two switches wired in series to give the circuit the "and" property will be called an "and" gate and will be shown by the symbol on the right.

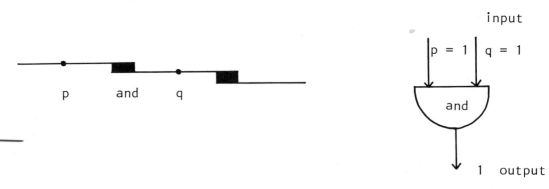

"and gate"

The two diagrams above have the same electrical property. Current only flows through the one on the left when both p and q are closed. Current only flows through the "and" gate when it is present at both points p and q. If 1 means the current is on and 0 means the current is off, then a "1" entering both p and q of an "and" gate has an output of 1 for the gate. If either one or both of the entering numbers at p and q is a 0, then the current does not flow through the gate. Which of the "and" gates below will have a 1 at the output connection?

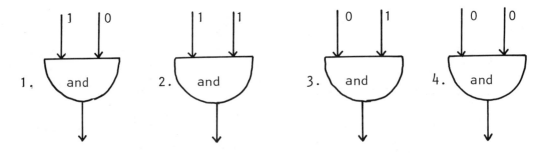

Only gate number 2 will have a 1 at its output connection as it is the only gate with a 1 at the left input and a 1 at the right input connection.

An "or" gate allows current to flow if either of its switches is closed because the switches are wired in parallel. Consequently, an "or" gate will have a 1 at its output connection if there is a 1 at either the left or the right input connection. An "or" gate and its equivalent symbol are shown on page 23.

Computers   23

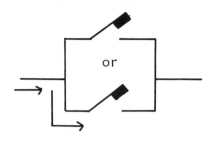 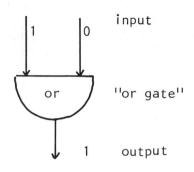

1 means the current is on and 0 means the current is off. Which of the "or" gates below will have a 1 at its output connection (the current flowed through the gate)?

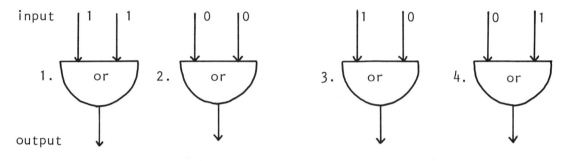

Since an "or" statement is false only when both of its parts are false, there will be a 0 at the output connection only when there is a 0 at both the input connections. Therefore, each of the gates above will output a 1 except gate 2.

There is only one other point to make to enable discussion of the circuit that can do addition of two one-digit binary numerals.

A "not-gate" (inverse) changes the condition in a wire to its opposite. When a 1 enters a not gate a 0 leaves the output connection and when a 0 enters a not gate a 1 leaves the output connection. The symbol for a not gate is shown below.

A look at the results desired when adding two one-digit binary numerals may help with the design of the circuit that can do this addition. When adding 1 and 0 or 0 and 1, the 1 should appear as the one's digit. When adding 0 and 0, a 0 should appear as the one's digit. But when adding 1 and 1 the answer

24   Introductory College Mathematics

is 10. Therefore, in this last case, the one's digit in the
answer should be zero. In other words when there is a 1 at
the first input and a 0 as the second input, the one's digit
in the answer should be a 1. When there is a 1 at the first
input <u>and</u> a 1 at the second input, the one's digit of the
answer should be a zero and <u>not</u> a 1. Notice how the "and,"
"or," and "not" gates are arranged in the diagram below to
produce a zero for the one's digit if two one's are added and
to produce a zero for the two's digit if a one and zero, or a
zero and one are added.

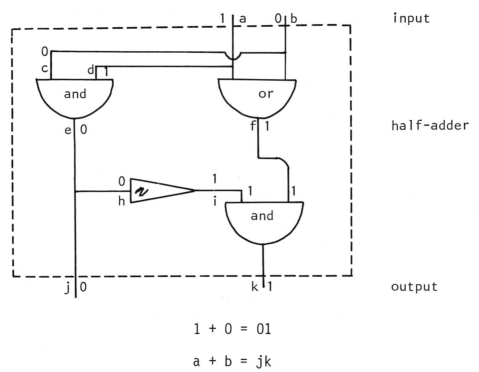

$$1 + 0 = 01$$

$$a + b = jk$$

The above circuit is called a half-adder and will add two one-
digit binary numerals. The diagram shows the condition of the
half-adder at the different points of the circuit when the inputs
are 1 and 0. Notice that when 1 and 0 are put in at a and b, the
result, 01, is shown at the output j and k. When 1 and 0 are
added in the machine we want a 0 for the two's digit and a 1 for
the unit's digit. When the 1 is entered at point a, a 1 goes
through the "or" gate at f and down to the "and" gate. Mean-
while the 1 from a is also routed to point d which is an "and"
gate. The 0 at b is routed to c. 1 and 0 give 0 at e and h.
But the 0 at h turns to 1 because of its passage through the
"not" gate. This produces a 1 at i which enters the "and" gate
with the 1 from f. 1 and 1 at the inputs of the "and" gate
produce a 1. Consequently, the unit's digit output is a 1,
and the two's digit output is a 0.

The diagram below shows the condition at the various points in the circuit for the addition of 0 + 1 = 1. Notice that the input of 0 at a and 1 at b is routed to both the and gate and the or gate at the top of the diagram.

0 + 1 = 1

a + b = jk

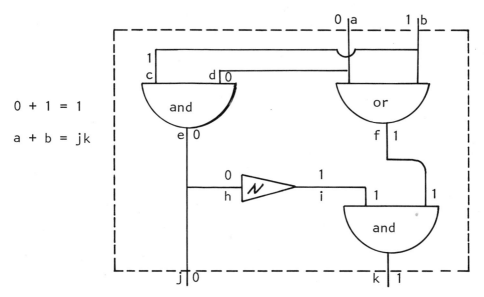

When the input at a and b is 1 and 1, as shown in the diagram below, the two 1's are routed to the and and or gates which both output a 1. The not gate changes the 1 from e to a 0 at i. The 0 at i and 1 at f produce a 0 at k, because of the and gate. The two's digit is a 1 because the 1 from e goes to j. Consequently, when 1 is the input at both a and b, the output at j and k is 10 and $1 + 1 = 10_2$.

$1 + 1 = 10_2$

a + b = jk

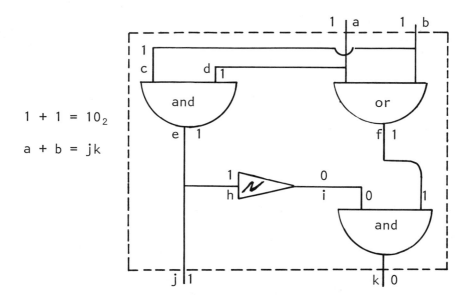

26   Introductory College Mathematics

Progress Test 4

Give the condition, 1 or 0 at points e, f, j, and k, for the following inputs.

	a	b	e	f	j	k
1.	1	0	__	__	__	__
2.	0	1	__	__	__	__
3.	0	0	__	__	__	__

ADDING THREE ONE-DIGIT BINARY NUMERALS

Two half-adders make up a full-adder and will add three one-digit binary numerals. Thinking about what has to be done when adding three one-digit numerals will help proper arrangement of two half-adders for this situation.

$$(1 + 1) + 1 = ?$$

$$10_2 + 1 = 11_2$$

Notice that in the problem above, the first two numerals are added to give 10_2. This addition can be done with a half-adder. If the result of the addition of the first two numerals is a two-digit numeral such as 10_2 another half-adder can add the one's digit (0) to the third numeral in the problem. When the sum of the first two numerals is 10_2 there will be a 1 to "carry over" to the second column of the problem. The diagram on page 27 shows how to arrange two half-adders to make up a full-adder which will add three one-digit binary numerals.

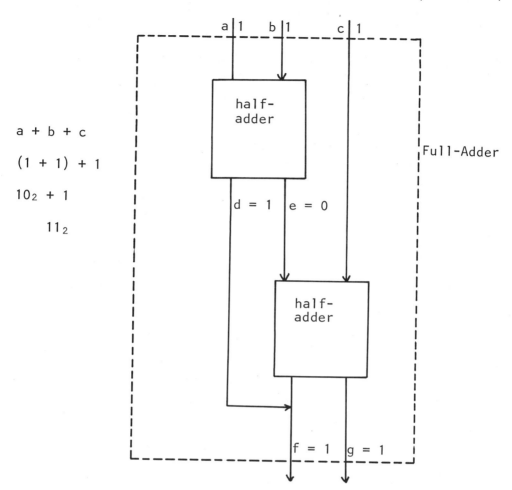

The inputs for adding three one-digit numerals are at a, b, and c. f and g are the output connections. When 1 + 1 + 1 is added, the previous figure shows the condition of the circuit at a, b, c, d, e, f, and g. For the top half-adder, the inputs are 1 and 1 and the output is 1 and 0 at d and e. For the lower half-adder the inputs are 0 and 1 and the output 0 and 1, yet at f and g there are 1's because the output d is connected to the output of the second half-adder at f.

Since the addition of binary numerals never involves numerals larger than 1, full and half-adders will handle the addition of any set of binary numerals. If the numerals to be added contain many digits, more half or full-adders will be needed to solve the problem.

28   Introductory College Mathematics

Progress Test 5

Using the full-adder diagram, use 0 and 1 to indicate the state of the current at the points given below.

	a	+	b	+	c	=	f g,	d,	e
1.	1		1		1		___	___	___
2.	1		1		0		___	___	___
3.	1		0		1		___	___	___
4.	1		0		0		___	___	___

Exercise Set 2

I. 1. True or false. A switch in an electric wire is analogous to a gate in a fence.

2. Will the current flow through the following circuits? (yes or no)

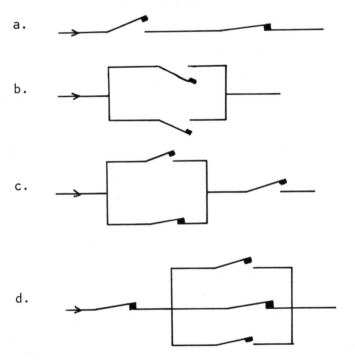

a.

b.

c.

d.

e.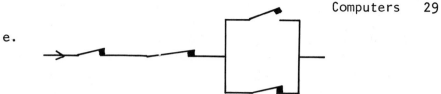

f.

3. Draw switching circuits to illustrate the following statements. Indicate whether or not the current will flow through the circuit.

   a. First switch closed and second switch closed.

   b. First switch closed or second switch closed.

   c. First switch closed or second switch open.

   d. p and q and r.

   e. p and -q.

   f. Either p or -q and then r.

   g. p and q and either -p or -q.

   h. either p or -q.

4. Use the formula $P = 2\ell + 2w$. If "p" means multiply ℓ by 2, "q" means multiply w by 2, and "r" means to add, draw a sketch that will give P if ℓ and w are entered in the computer.

5. Add the base two numerals below. Use base two numerals.

   a. $1_2 + 1_2$ b. $1_2 + 1_2 + 1_2$

   c. $11_2 + 10_2$ d. $101_2 + 111_2$

   e. $1_2 + 0 + 1_2$ f. $1010_2 + 101_2$

   g. $1111_2 + 1_2$ h. $1_2 + 11_2$

30   Introductory College Mathematics

6. Using the full-adder diagram, use 1 or 0 to indicate the state of the current at the circuit points given below.

	a	+	b	+	c	=	fg,	d,	e
a.	0		1		1		___	___	___
b.	0		1		0		___	___	___
c.	0		0		1		___	___	___
d.	0		0		0		___	___	___

II. Challenge Problems

1. Make a switching circuit to show: (either p or -q and either q or -p). Will current flow through the circuit?

2. Arrange 1 full-adder and 2 half-adders for the addition of four one-digit binary numerals.

3. Arrange three full-adders and one half-adder for the addition of two four-digit binary numerals.

STORED PROGRAMS, TRANSISTORS, AND OTHER DEVELOPMENTS

Dr. John von Neumann, considered by many as the most creative scientist alive in the 1940's and 1950's, conceived the idea of storing the instruction program in the memory of the computer, an action which would eliminate the time required to wire a program into a computer. Von Neumann, a Princeton mathematician, wrote: "Since the orders that exercised the entire control are in the memory -- the machine can extract numbers from the memory and replace them -- indeed this is its normal mode of operation. Hence, it can change the orders (since these are in the memory!) -- the very orders that control its operations." One copy of the first stored program computer was called Johnniac over von Neumann's objections.

Both ENIAC and the first stored program computers following ENIAC were hastened in their inception by the needs of war. The scientists involved in weaponry and in the atomic bomb

developments could not handle all the arithmetic necessary to
solve their problems. As von Neumann said of one problem re-
lating to nuclear weapon design, "Probably in its execution,
we shall have to perform more elementary arithmetical steps
than the total in all the computations done by the human race
heretofore." Because of the problems to solve, the U. S. Army
sponsored research on electronic computers. ENIAC was the first
and the largest electronic computer to be produced under their
sponsorship. It occupied about 1800 square feet, which is the
area of a house 30 feet wide and 60 feet long. Newer computers
have a larger capacity while occupying less space, due largely
to the advent of the transistor. Transistors are so small that
a magnifying glass is needed to study them. Transistors also
are much faster than vacuum tubes, operating in billionths of
a second instead of millionths of a second.

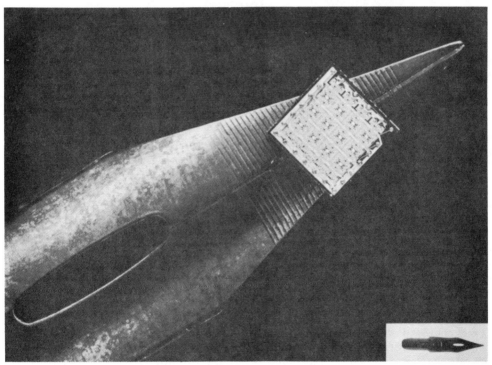

A monolithic circuit on a pen nib.

Another advantage transistors have over tubes is the fact that
they produce no operating heat. A vacuum tube must be hot to
function properly. When fifteen to twenty thousand tubes are
used in a computer an enormous cooling problem is encountered.

32   Introductory College Mathematics

Transistors produce very little heat, which helps extend their life and that of the other computer parts. In fact, tests indicate that transistors have an almost unlimited life. Consequently, thousands of consecutive operations can be carried out by a transistorized computer with virtually no error. Like the vacuum tube and relay, the transistor gives the computer the power to make choices between two numbers or instructions. The 1956 Nobel Prize was awarded to J. Bardeen, W. H. Brattain and W. Shockley of Bell Telephone Laboratories for their work in inventing the transistor in 1948.

Monolithic circuits used in IBM System/370.

Since the invention of the transistor, computers have become increasingly sophisticated. Using micro-technology, whole sections of the computer containing many transistors and other components are embedded in a plastic chip so small that 50,000 chips can fit in a thimble. As a result, computers that used to fill a room are reduced to the size of a small box and can do arithmetic operations in a few nanO-seconds (billionths of a second.) Memory units that used to be networks of wires intersecting in magnetic doughnut-like cores, one for each digit in an instruction, now use interchangeable magnetic tapes or discs. The capacity of the memory, using tapes or discs, is only limited by the number of tapes or discs available.

A disk drive.

Input devices now do not need punched cards. Instead, the oddly shaped letters used on computer cards can be seen by a scanning sensor. Because of the increased efficiency of modern technology, the cost of 100,000 arithmetic operations has decreased from $450 in 1940 to 0.02¢ in 1975, making the use of computers feasible for practically every business and social agency. In fact, much business competition can only be met by companies with access to the symbol handling capabilities of a computer.

Computers have been used in a great variety of ways since their first commercial introduction in the early 1950's. The number of different computer applications is growing so rapidly that any list of computer uses is obsolete upon its publication. A partial list includes the following categories:

Filing and Information Retrieval -- Files in business and in government agencies can be held and controlled by a computer. Upon request, the computer can search its memory and produce the desired information. Personnel resource data, travel reservations, bank records, records and billings for businesses, credit checks, library processing and law enforcement files are examples of filing and information retrieval by computers.

Data Manipulation -- Data can be arranged and printed out in a variety of ways: graphs, scoreboards, matrix arrays and tables are some of the ways data can be arranged.

Simulation -- Models of systems such as transportation, war, flood control, automobile, political, human behavior, and the stock market can be simulated by computers to study the consequences of decisions made in the system.

34   Introductory College Mathematics

Computer in use at nursing station in hospital.

Control -- Processes in hospitals, factories, transportation lines, space travel, and printing plants can be controlled by computers with splitsecond speed. Speed is important between the observation, decision, and correction of a processing situation to insure effective control of the process.

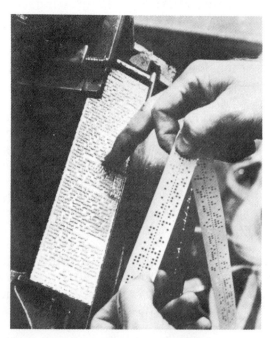
Typesetting by computer.

Air traffic control.

Pattern Recognition -- The computer can be used to recognize symbols, words, mail and patterns in scientific work in Zoology, Biology, and Engineering problems.

Calculations -- Any mathematical calculation from the simplest to the most complex can be accomplished by a computer.

Heuristic Problem Solving -- Some of the most novel and creative uses of computers are in this category which includes discovering proofs in mathematics, generating new moves in checkers and chess (There is international competition among chess-playing machines.) learning to deal with complex problems in social - economic or social - technical areas and oral interchange with the computer to solve problems.

Many computer applications overlap several categories. Computer assisted instruction, counseling, and dispatching are computer functions which use information retrieval, control and problem solving.

36   Introductory College Mathematics

In previous parts of this section, you have seen how computing devices have been spurred in their development by the needs of man. Computers have been powered by hand, by machine, by electro-mechanical devices, and, finally, by electronics. Fingers and rocks were originally used as input and output by man when computation was done. Later, numbers on a wheel, punched cards and tape discs, electric typewriters, and magnetic tape were used to record both the input and output of computers.

Methods of instructing a computer were originally in the head of primitive man. Now, they are stored on the computer memory and can be changed by the computer itself. This ability to change its program while following a program lends credence to myths of computer brainlike capacity, but the control is still in man's hands.

The computing machines mentioned previously probably seemed to possess mystical powers to the people of their particular time. Yet, as advanced as each new computer seems, it is soon out of date. Since there are so many more people working in computer design today compared to the number working in 1940, we will probably see an even greater change in computer uses and technology in the future.

Progress Test 6

Place the following computer applications in one of the appropriate categories in the list at the right.

APPLICATIONS	CATEGORIES
Weather plotting	Filing and Information Retrieval
Studying a design before an object is built	Data Manipulation
	Simulation
Studying a cultural situation and recommending changes	Control
	Pattern Recognition
Translating languages	Calculations
Sorting	Heuristic Problem Solving
Personnel files	

APPLICATIONS (continued)

Construction cost control

Machine control

Analysis of political systems

Airplane reservations

Scientific computations

Automatic washing machine

Computerized criminal history program

Games

WHAT IS A COMPUTER?

As indicated in the section on computer history, a computer consists of three basic parts: the input, the processor, and the output unit. As a human solves problems, the senses pick up the data to serve as the input, the brains process the information, and the eyes, mouth, fingers, and the rest of the body serve as output devices. Consequently, people and computers have much in common.

In most installations computers seem to be a set of boxes arranged about a room. (See illustration on page 38.) There does not appear to be any connection between the boxes. But they are wired together with the wiring hidden under the floor. A simple diagram of the basic computer parts is shown below.

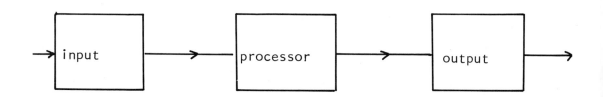

38   Introductory College Mathematics

A medium-size computer installation.

The computer can be thought of as three people who handle the
data as it enters the input, is processed, and is printed by
the output printer. The computer memory can be viewed as a
set of boxes each with a window so that the symbol inside may
be seen. The word "symbol" is used rather than "number" since
computers really are symbol processors. Some of the symbols
are numbers. As the computer is following a set of instructions
(algorithm) on a flow chart, a window box is associated with each
variable in the instructions. For example, the computer may be
figuring the payroll of a company using the formula

$$\text{Pay} = \text{Hours} \cdot \text{Rate}$$

In the formula above, pay, hours, and rate are called variables.
They are called variables because they will vary in value depending
on the situation. For example, if hours = 40 and rate =
$8.00 per hour, the pay equals 40 · $8 = 320. If the next person
worked 41 hours, and the rate remains at $8, then the pay
will be 41 · $8 = $328. The box in the memory associated with
hours will have a window so that the number of hours in the box

can be seen. Similarly, there will be a window box for rate and one for pay in the memory.

Using the window boxes to hold variables in the memory and three workers; a reader, an assigner, and a calculator, a model of a computer can be described. This model will demonstrate the action inside a computer as it produces answers in millionths of a second.

The reader, assigner, and the calculator each have a specific job. The reader's job is to go to the window boxes, read the numbers inside and tell the calculator. The assigner also works for the calculator. His job is to take the answers supplied by the calculator to the appropriate window box, dump out its contents and place the new number inside. The next paragraph shows how the three workers make up a payroll.

Suppose that a punched card showing Bill Koski's rate of pay and time enters the input unit of the machine. The assigner takes the data and places the appropriate numbers in the rate box and the time box. The window box labeled pay is either empty or contains the pay of the person that was figured previously. When the computer is signaled to start, the calculator looks at the flow chart which contains his instructions and finds that he needs to know the hours and rate for Bill. He sends the reader to the memory to read the contents of the "hours"

and "rate" window boxes. When the reader comes back and gives the calculator the information, the calculator, following the algorithm on the flow chart, multiplies the hours times the rate and gives the answer to the assigner. He tells the assigner to go to the pay window box in the memory, dump out the contents, and insert the new pay number. The output unit will then print Bill's pay on a check payroll, or tape as prescribed by the flowchart (program).

The previous paragraph described one cycle on the computation of a payroll. As the next punched card enters the input unit, the assigner puts the hours and rate in the correct window boxes in the memory, the calculator sends the reader to the memory to bring back the data, the calculator figures the pay and sends the assigner to place the answer in the pay window box. When all the cards have been processed, the machine can be stopped by a card punched "stop." Of course a computer is more sophisticated than the example just given, but the functions of reading, calculating, and assigning are basic to all computers.

Describing a computer as consisting of three units: input, processor and output, does not indicate the complex role of the processor. Actually, the processor can be thought of as three units itself: memory, arithmetic logic, and control. They can be interconnected as shown in the following diagram. Notice that the control unit is connected to all the other units.

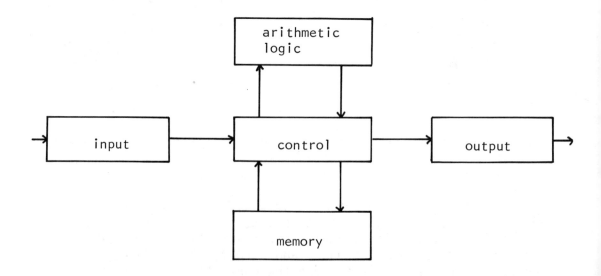

Progress Test 7

1. A cave man's fingers serve as input, output and memory of his computer. What served as his control unit?

2. Which of the features at the right are possessed by:

   a. Mechanical calculator
   b. An abacus
   c. Pascal's machine
   d. Leibniz' invention
   e. Hand-held electronic calculator

   a. Memory capacity
   b. Input capacity
   c. Control
   d. Output capacity
   e. Arithmetic unit

FLOWCHARTS AND ALGORITHMS

To use a computer to determine the correct shape of an airplane wing or to compute the phone bills for a city requires thousands of distinct operations that must be done in proper order. The person who plans these operations is called a programmer.

The rule determining the computer's operation is called an algorithm. The programmer lays out the algorithm on a flowchart. A flowchart is a diagram that shows the path of the data through the algorithm.

Naturally, the programmer requires a thorough knowledge of the problem, the computer, and how to communicate with it.

An algorithm tells how to carry out a process. Some examples of algorithms are recipes, instructions on how to assemble a toy, and the rule on how to find the square root of a number.

The following algorithm is a list of instructions for frying an egg.

1. Put a frying pan on the stove.

2. Turn on the heat.

3. Put grease in the pan.

4. Break the egg into the pan.

5. Cook the egg.

6. Take the egg out of the pan.

7. Turn off the stove.

A flowchart showing the above algorithm is shown below.

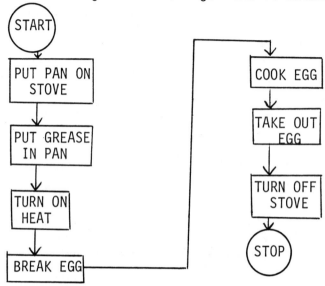

The diagram shows some of the characteristics of a flowchart.

1. It contains a series of commands or directions placed in polygons.

2. At the beginning and end there are the directions to START and STOP set off in a circle.

3. The arrows between the parts of the chart are used to indicate the order in which the directions are to be followed.

A flowchart is a great aid to a programmer, as it provides a graphic view of all the steps in an algorithm. It is as useful to the programmer as a map would be to a tourist.

You have probably noticed that the flowchart given previously could have many improvements: the temperature of the pan, the size of the egg, the hardness of the egg, whether it should be turned over, and when to turn off the heat. All of the instructions for handling these situations could be included in the flowchart.

The chart below shows how to provide for the cooking time of the egg. A decision on the condition of the egg could be made by inserting a decision box in the program to handle this situation. A decision box many times asks a question and requires a yes-no answer. Decision questions are indicated by a diamond box. Decision boxes may also contain true-false statements.

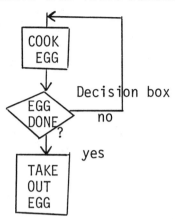

The loop back to "cook the egg" is called a "feedback loop". Feedback loops are used in programs to shorten the program. Otherwise a branch of the program would have to be diagrammed to handle the alternate situation where the egg is not done.

Sometimes the same operation needs to be done many times. For instance, we might want to fry 600 eggs. 600 boxes like those below would be inserted in the program to accomplish this.

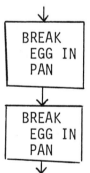

But, an easy way to accomplish the same result is to put in a decision box with a feedback loop as shown on page 44.

44 Introductory College Mathematics

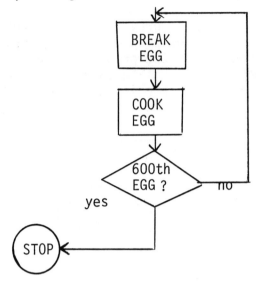

The flowchart below shows an improved version of the frying-egg flowchart given previously.

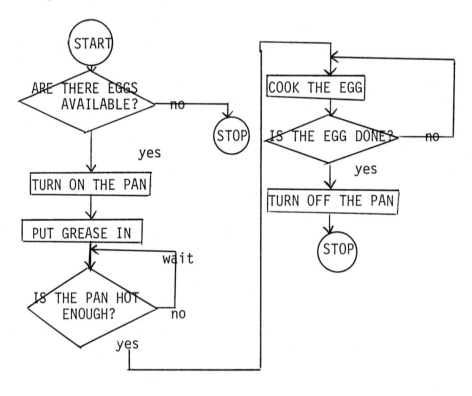

Although the chart on page 44 allows for more control over the process than the first flowchart, there is at least one glaring error in the logic of the chart. There is no instruction to break an egg into the pan! To prevent such errors, the flow of the data through the chart has to be carefully studied to try to foresee all situations that could arise. Even then, programs may have to be run many times before they are completely debugged.

Progress Test 8

1. Flowchart the following:

    a. Putting on a shoe.

    b. Putting on a pair of shoes.

    c. Sharpening a pencil.

FLOWCHARTS EXPANDED

One of the strengths of computers is their ability to do routine operations endlessly and accurately. This characteristic enables computers to find solutions for problems that humans could do if they lived long enough. For example, the area of the shaded section under the curve below could be figured very accurately if it was divided into 1,000,000 rectangles. The area of each rectangle could easily be figured and added to all the other areas to give a very close approximation to the shaded area. The time required to compute 1,000,000 areas and find their sum would be lengthy for humans, but not for computers as computers can compute in billionths of a second.

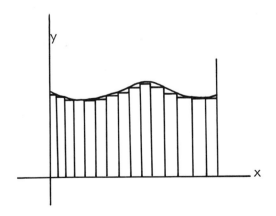

46 Introductory College Mathematics

Another example of a problem that is simple but time consuming if done without the aid of a computer, is the following:

The pay for each of 35,000 employees of a large corporation could be quickly computed and printed out by programming a computer to follow the instructions in the flowchart below. It is assumed that a punched card has been made for each of the 35,000 people showing their rate of pay and the hours worked.

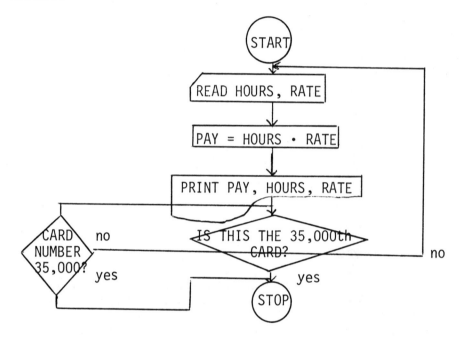

The flowchart preceding this statement and those following will follow generally accepted flowchart notation used by computer customers.

In the previous flowchart, different types of instructions are indicated by different symbols. These statements and their corresponding symbols are listed below.

START AND STOP statements are in circles

READ statements are in figures that look like IBM cards.

Assignment statements are in
rectangles. In this statement,
X is assigned the value (2x + 3).

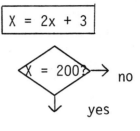

Decision statements are in
diamond boxes.

PRINT statements are written
in figures that look like torn
printout paper.

The payroll flowchart tells the computer to do the following:

1. Read the rate and time of each card and hold all readings in memory.

2. Multiply each rate by its time and assign the product to PAY. Assignment statements almost always contain an equals sign.

3. Print out the pay, rate, and time with the computer's output typewriter.

4. If the last card read is not the 35,000th card, the decision box instructs the computer to read the next card. If it is the 35,000th card, the computer is told to STOP.

The reader may be wondering how the computer knows that the 35,000th card has entered the machine. The next flowchart shows how data cards can be counted as a computer is following a process.

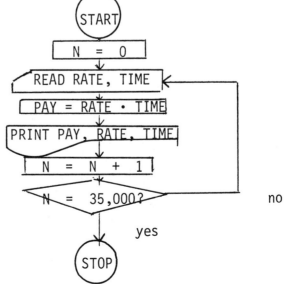

The previous flowchart has two or more assignment boxes and a change in the statement in the decision box compared to the one on page 46. The assignment statements, N = 0 and N = $\underline{N+1}$, enable the computer to count the cards as they enter the machine. When the first card enters the input unit the number 0 is assigned to N. After the PRINT statement, the assignment statement N = N + 1 adds 1 to the value of N and erases the previous number that was assigned to N. Remember that an assignment is destructive. That is, every time a number is placed in a window box in memory, the previous contents of the box are dumped out. The statement N = N + 1 raises the value of N by 1 and assigns this new value to N. Consequently, every time a card goes through the machine the value of N increases by 1. The old value of N is destroyed.

The statement in the decision box, N = 35,000? is shown in the way statements are usually found in decision boxes. It is a statement whose truth is questioned. In this case, if it is true, the process stops and if it is false, the next card is read.

The next example shows a flowchart for the process required to solve ninety-five different quadratic equations: a problem that might occur in a scientific situation. $ax^2 + bx + c = 0$ represents a typical quadratic equation and it is assumed that the numbers a, b, and c have previously punched on 95 data cards ready to be processed by the computer.

The following statements explain the steps in the flowchart. N = 0 and N = N + 1 are assignment statements that have a counting function in the process. That is, as the data cards are run, N will equal the number of cards run because each time the loop is run N is increased by one.

"Read a, b, c" means that the numbers representing a, b, and c are placed in the memory and the previous values of a, b, and c are erased from memory.

"$x = \frac{-b + \sqrt{b^2 - 4ac}}{2a}$" assigns the value of the right side of the equation to x. In other words, using the numbers for a, b, and c the computer figures the value of the right portion of the equation and assigns that number to x.

"PRINT x" directs the output unit of the computer to print the value of x on the printout sheet.

Computers 49

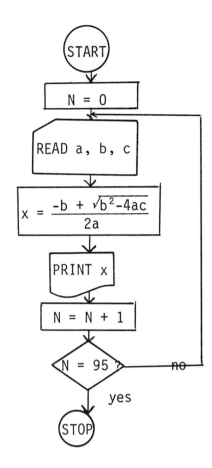

"N = N + 1" raises the value of N by one and assigns the new value to N while destroying the old N value. When the N = 95 the decision box directs the computer to stop the process.

What will be printed out on the printout sheet? Since the only print direction involves x, the printout sheet will be a list of 95 values of x: one for each of the 95 equations.

---

50   Introductory College Mathematics

Progress Test 9

Name the process shown by the following flowcharts:

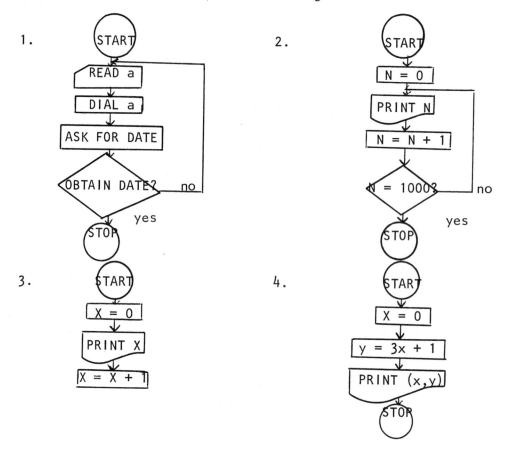

Another common problem well adapted to the computer's tireless ability to handle symbols in finding the average of a large set of numbers. Suppose, for example, a flowchart is needed to average the eleventh grade test scores for all the eleventh graders in New York City. To obtain a hint on the instructions needed, the step necessary for a paper and pencil solution can be thought through.

First, the first two numbers would be added. Then to that sum the next number would be added, etc. until the last number is added to the sum. Then the sum would be divided by the total number of grades to obtain the average. The sum will be zero at first, before any grade is added, and as each successive data card goes through the machine, the sum will increase.

Also, the number of grades will have to be counted so that the total number can be used as a divisor when the average is computed.

Consequently, the assignment statement SUM = 0 is needed at the start to give the sum an initial value of zero. Then as each grade is processed it will have to be added to the sum giving a new value to the sum. The statement SUM = SUM + GRADE will accomplish this. Remember the old sum value will be destroyed as the new value is assigned to SUM. An algebra student will probably question a statement such as SUM = SUM + GRADE but in computer programming it is legitimate because assignment of numbers to memory always dumps out the value already in the memory.

Study the next flowchart carefully to determine if it will find the average of 11,278 grades.

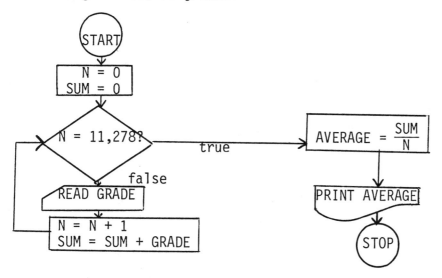

Notice that this chart directs the computer to follow the instructions in the loop until N = 11,278. Then it is to figure the average and print it out. The statements, N = 0 and SUM = 0 have to be made before the loop so that the values of N and SUM will keep increasing. If they were inside the loop they would revert back to zero everytime a grade was processed and the number of cards would be lost.

The decision box in the previous example depended upon knowing that there were 11,278 grades to average. Can a flowchart be made that will stop the looping action and figure the average without knowing how many grades to expect? After all, counting the grades when they number in the thousands can be expensive. The answer to the previous question is yes. A description of the process involved is given in the next paragraph.

52 Introductory College Mathematics

Suppose the highest grade possible on the test is 500. Then a false grade such as 1000 can be punched on a data card which is placed at the end of the deck of data cards. The 1000 card at the end of the deck, together with an appropriately worded decision statement will direct the computer to average the grades even though the computer will have to count them to find their total number. The flowchart has to be constructed so that the false grade of 1000 will not be averaged with the rest of the grades, but at the same time cause the computer to give the desired result. The process shown in the following flowchart will average the grades even though their total number was not previously known.

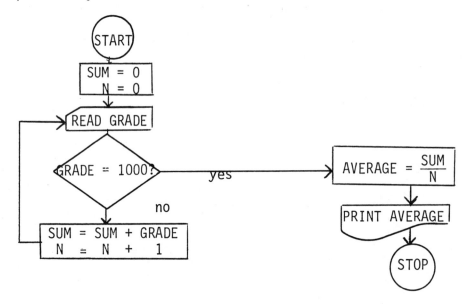

Notice that in the flowchart above, the decision box is placed ahead of the assignment statement SUM = SUM + GRADE so that 1000 will not be added to the sum of the grades and affect the average of the grades. The statements N = 0 and N = N + 1 are needed to instruct the computer to count the grades.

Progress Test 10

1. Suppose the shaded area on the right is subdivided into 2,000 rectangles each 0.01 inch wide and the length of each rectangle has been figured and punched

on data cards. Using the formula A = LW, write the statement that will assign the area of each rectangle to a letter A. Let L represent the length of each rectangle.

2. What value would first be assigned to SUM if the process is to find the sum of all the 2000 rectangle areas?

3. What is the next assignment statement involving SUM?

4. Write the two assignment statements that will direct the computer to count the number of areas?

5. Make a flowchart showing the process that will figure the area of each of the rectangles and compute the sum of all the areas.

Exercise Set 3

1. Make a flowchart for each of the following algorithms:

   1. a. start, b. select shirt, c. put left arm in left sleeve, d. put right arm in right sleeve, e. button shirt, f. stop.

   2. a. start, b. take out spare tire, c. take out jack, d. take off hub cap, e. loosen nuts, f. jack up car, g. take off nuts, h. take off tire, i. put spare tire, j. tighten nuts, k. replace hub cap, l. lower jack, m. put tools in car, n. stop.

   3. a. start, b. select tie, c. tie the knot, d. if tie length is not correct go to b, e. stop.

   4. a. start, b. if there is no hamburger available go to j, c. turn on pan, d. wait 60 seconds, e. if the pan is not hot enough go to d, f. fry hamburger, g. wait 30 seconds, h. if hamburger is not done go to g, i. take out hamburger, j. stop.

   5. a. start, b. check pencil point, c. if pencil is sharp go to h, d. place pencil in sharpener, e. turn crank, f. check pencil, g. if pencil is not sharp go to d, h. stop.

54 Introductory College Mathematics

6. a. start, b. read A and B, c. if A $\leq$ B go to b,
    d. print A, B, e. if there are more pairs A, B go
    to b, f. stop.

7. a. start, b. read L, W, c. let P = 2L + 2W,
    d. print P, L, W, e. if there are any more pairs
    L, W go to b, f. stop.

8. a. start, b. let N = 0, c. read R, T, d. let
    PAY = R·T, d. let N = N + 1, f. print PAY, R, T,
    g. if N $\leq$ 10 go to c, h. stop.

Describe the process shown by each flowchart

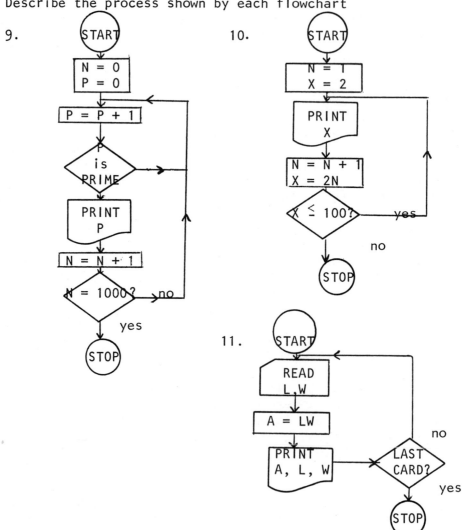

12.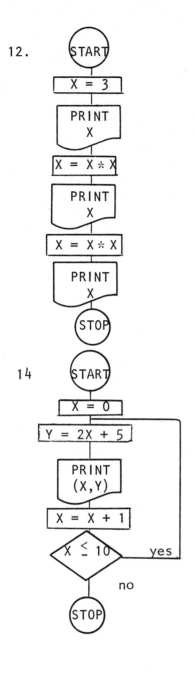

13.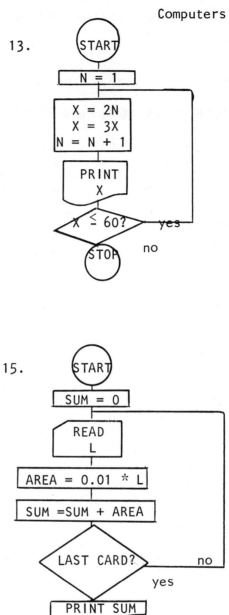

14.

15.

COMPUTER LANGUAGES (HOW TO TALK TO A COMPUTER)

After a programmer has constructed and revised a flowchart until it's thought to be feasible, the flowchart needs to be communicated to the computer. How can the programmer make the computer follow the flowchart's instructions? After all, the computer only responds to electric impulses. The answer to this question has changed over the years as computer technology has become more sophisticated. At first, people like Howard Aiken, who developed the first electro-mechanical computer, meticulously developed a flowchart for the problem he was attempting to solve. The flowchart was then written into a program. Every operation has to be considered and included in the program. As data entered the machine, each item had to be assigned an address in the memory. The assignment of numbers of instructions to an address in the memory had to be inserted in the correct place in the program. Then, when a number at some address in the memory was needed for a certain step in the process, the instruction to read that number had to be included in the program. Because of these and other details to consider, early programmers often spent days developing a program. Then, when they were satisfied with the program, the instructions were wired into the computer. That is, certain connections in the computer had to be made so that the desired action would result. If a decision box was in the flowchart, the wiring necessary to produce that action in the machine had to be included in the wiring. Naturally, the people wiring in the program had to be highly skilled in several areas. Because of the time and cost required to properly program a computer, efforts were made to simplify the process.

One step in the simplifying process was to prewire every conceivable action that could occur in a flowchart into the computer so that each action could be activated by a signal from a punched card. Each action was assigned a number so that when a card with a particular number entered the machine, the corresponding action was taken by the machine. With this technique, a programmer arranged the instruction cards in an order to correspond with the instructions on the flowchart. For example, if the number 203 on a card meant to read a number in address 203 in the memory, when the computer came to that particular card in the program deck, it read the number on address 203. Each program was represented by a stack of cards which could be run again whenever that program was desired. Then when the data was available on cards, magnetic tape or discs, the stack of program cards was inserted into the computer and the program was run.

The programming language described above is called machine language. It consists of numbers that activate the computer. Machine language is the most primitive language used to instruct a computer and consists of about 200 coded operations. A typical code number for an instruction is given in six digits. The first three digits indicate an operation to be done. The second three digits show the appropriate address in the computer memory. The following is a description of how the simple operation of adding two numbers would be carried out in machine language. Suppose the numbers to be added are carried in addresses 201 and 202 of the computer memory. The first step in the addition is to take the number in address 201 and deposit it in the accumulator. The designation of the operation is +500 201 in which +500 stands for clearing the accumulator and depositing in it the number stored in address 201. The "sign" of the instruction was important as -500 indicated an entirely different operation. Then came the message +400 202 which meant to add the number stored in address 202 of the memory to whatever is in the accumulator. The next step is to take the sum and store it at some address in the memory until needed in some other part of the computation. As someone said, "A computer is an idiot; it has to be told every minute step in a process."

Developing languages that would be translated by the computer itself into machine language was the next step in the development of computer communication. Because it was difficult to conceive that a computer could translate a program into machine language as efficiently as a human, the development of such languages was slow. However, computer languages have changed so much that it is now almost possible to draw up a program in English, and let the computer translate it into machine language saving hours and even days of translation time.

Some of the translation languages that were invented to bridge the gap between English and machine language are ALBOL, COBOL, FORTRAN, PL1 and BASIC. Every statement on FORTRAN, for example, is translated by the computer into from four to twenty machine language instructions. Tremendous effort was required by those men who designed and perfected FORTRAN. First, machine language programs had to be analyzed to find the basic processes that were frequently used in the programs. These were given FORTRAN names. Once the language was perfected, the programmer could forget machine language altogether and use the FORTRAN statements altogether. At first, it was thought that the FORTRAN program would contain more steps in machine language than one constructed by a human programmer. As the languages were perfected, programmers would race a FORTRAN program against one made by a man to see

58   Introductory College Mathematics

which was most efficient. After some refinement, FORTRAN programs were just as efficient as those made by humans.

The following section of a machine language program shows the detail required when writing a machine language. The expression b^2-4ac is being evaluated.

003 04060	Multiply number in arithmetic by number in 60
004 06070	Place number in arithmetic in 070 for storage.
	Clears arithmetic
005 01050	Register b in arithmetic
006 04051	Multiplies b in arithmetic by b in storage, now b is in arithmetic.
007 03070	Subtract 4ac from b^2 (4ac was computed previously)
008 06071	Places b^2 - 4ac in 071 storage and clears arithmetic

The advantage of FORTRAN can be seen by one example. After a one-day course in FORTRAN, a man programmed a process in four hours using 47 FORTRAN statements. These were compiled (translated) by a 704 computer into 1000 machine language instructions in six minutes. It was estimated that probably three days would have been needed to write the program by hand in machine language and an unknown time would have been needed to de-bug the program. The hand written program would not have been any faster than the FORTRAN program.

John Backus and Irving Ziller enjoyed some amusing moments while doing the early development work on FORTRAN. They remembered that they would put in a FORTRAN program and then go to the output printer to see the result. They could remember saying "Look what it did now." Their results were frequently different from what they expected because the full implications of their instructions were not understood. Many times there were machine language instructions that would not come from any FORTRAN statements and some FORTRAN statements that would not produce machine language translations.

Computers   59

The relation between flowcharts, translation languages, machine language and the computers is indicated by the chart below.

FLOWCHARTS AND BASIC LANGUAGE

The goal of this section is to introduce some of the symbolism of BASIC. The goal does not include teaching a complete course in BASIC Programming.

The statements on the left in the display below are the BASIC language statements for the corresponding statements of the flowchart. All BASIC language is written using capital letters.

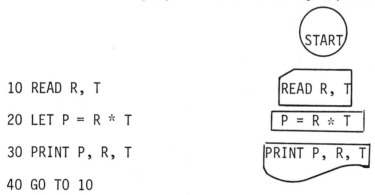

10 READ R, T

20 LET P = R * T

30 PRINT P, R, T

40 GO TO 10

There are several lessons to be learned from the BASIC statements above.

1. The numbers 10, 20, 30, 40 are used to number the statements. Multiples of 10 are used for this purpose to provide room between the statements should other statements be needed between the present statements.

2. BASIC statements are directly related to natural English usage. Other translation languages are not as easily read by the untrained person.

3. Statements starting with LET are assignment statements. LET statements assign a value to a variable and always consist of variables, numbers and BASIC symbols. For example, if SUM = SUM + GRADE is an assignment statement in a flowchart, the corresponding LET statement is LET S = S + G. The following LET statement is incorrect because words, instead of variables, are used: LET SUM = SUM + GRADE.

4. GO TO statements are used when it is desired to loop back to a previous statement.

5. There is a close relationship between BASIC language and flowcharting.

The statements above do not constitute a complete program. There is no data or stop statement included.

The program below is a complete program including the data and printout of the results.

```
10 REM PAYRØLL
20 READ R, T
30 LET P = R * T
40 PRINT R, T, P
50 GØ TØ 20
60 DATA 7, 40, 6.50, 32, 7.50, 30
70 END
```

RUN

PAYRL	18:28	THURS	04/4/74
7	40		280.00
6.50	32		208.00
7.50	30		225.00

ØUT ØF DATA IN 20

What do some of the added statements mean? Statement 10 is a remark on the nature of the program. It identifies the program. Statement 60 is the DATA INPUT statement. The data is arranged in pairs to correspond to the variables in the READ statement and the LET statement. The first pair is 7, 40. Consequently, 7 represents rate of pay and 40 is the time for this employee. DATA INPUT figures will always correspond to the variables in the READ instruction. When the data has all been processed, the computer will stop and print OUT OF DATA and give the READ statement number. The lower part of the display on page 60 is the printout of the results of running this program. It is a direct result of the PRINT statement in the program. The left column contains the rate, the center the time, and the right column the pay because the PRINT statement said PRINT R, T, P.

The table below lists the BASIC algebra symbols and the ordinary algebra equivalent symbol.

BASIC expression	Ordinary Algebra Expression
X + Y	$x + y$
X - Y	$x - y$
X * Y	xy
X / Y	$\frac{x}{y}$
X ↑ 2	x^2

Notice that the symbols / and ↑ enable the expressions to be typed on one line instead of using two as required by the expression $\frac{x}{y}$. The next example of a program written in BASIC shows how a program can generate its own data and illustrate the use of an IF statement. Notice that the data generating statements are similar to the counting statements.

62 Introductory College Mathematics

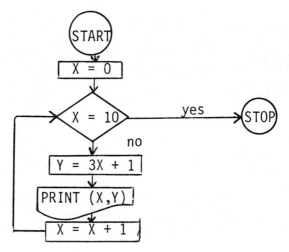

```
10  REM FINDING ØRDERED PAIR SØLUTIØNS

20  REM OF LINEAR EQUATIØN

30  LET X = 0

40  IF X = 10 THEN 90

50  LET Y = 3X + 1

60  PRINT (X,Y)

70  LET X = X + 1

80  GØ TØ 40

90  END
```

IF statements are used to enable the computer to make decisions. Statement 40 above tells the computer to go to 90, the END, when X = 10. If x is not equal to 10 direction 50 is followed and then 60 and 70. Since X = 0 the first time the process is followed, the first item printed out is the ordered pair (0,1). Then, the assignment statement X = X + 1 raises the value of X to 1 as the computer follows the loop. What will be the next item printed out? Since X = 1 and Y is assigned the value of 3X + 1, the next ordered pair to be printed is (1,4). As the loop is followed, X takes on the values 2,3,4,5,6,7,8, and 9 while Y is assigned the value of 3X + 1. Therefore, the rest of the ordered pairs printed out are (2,7), (3,10), (4,13), (5,16), (6,19), (7,22), (8,25), (9,28). Then, when X = 10 the IF statement instructs the computer to go to 90 which stops the machine.

For another example of BASIC programming, the following flowchart and program are provided. Notice the READ instruction and the loop is accomplished in the program.

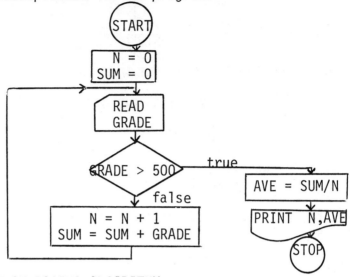

```
10    REM AVERAGING ALGØRITHM

20    LET N = 0

30    LET S = 0

40    READ

50    IF G > 500 THEN 90

60    LET N = N + 1

70    LET S = S + G

80    GØ TØ 40

90    LET A = S / N

100   PRINT N,A

110   END
```

The IF statement above agreed, in sign, with the statement in the decision box of the flowchart. Sometimes, to facilitate programming, an IF statement may be the opposite of the decision statement in a flowchart. For example, the decision box may be:

64   Introductory College Mathematics

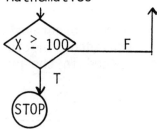

The corresponding IF statement will be:

    80   IF X < 100 THEN 20

    90   END

The sign "≥" was reversed to "<" so that a straightforward IF statement could be worded.

Study the previous flowcharts and programs and take the following Progress Test.

Progress Test 11

For each flowchart, write a program in BASIC.

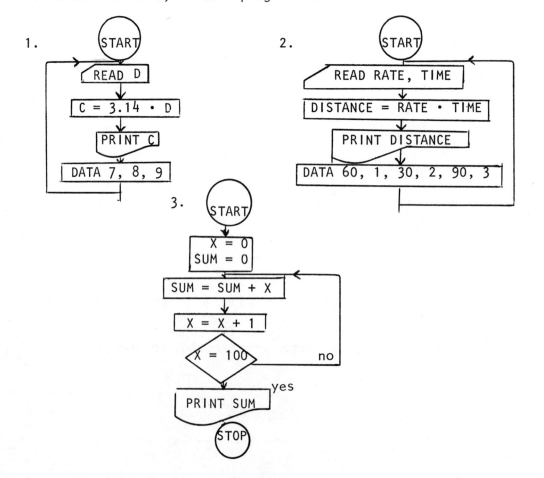

BASIC, FORTRAN and other programming languages instruct computers' to read and execute algorithms that are described by flowcharts. BASIC was originally developed by John G. Kemeny and Thomas E. Kurtz at Dartmouth College for use by beginners in programming. It is an all purpose instruction code that can describe a wide variety of algorithms. Since many beginners first contact a computer through a teletype machine wired to a computer which may be thousands of miles away, the characters used in BASIC are those found on any teletypewriter.

BASIC is usually used with a time-sharing system. Many customers share a computer's time by telephone connection with a computer and communicate with it using BASIC. Usually each customer is hardly aware that he is sharing the computer's time because the machine computes so quickly that it is rare for a customer to wait for the computer to run a program. If the computer is busy with another program when contacted by a person at a terminal, it will type out "WAIT" and a few seconds later it will have finished with the other program and be ready for action. The customer types in the program and data, and the computer calculates the answer typing it on the same teletypewriter.

The following statements summarize the procedure when people use a computer to solve a problem.

Step 1. After studying a problem thoroughly the programmer designs a flowchart that hopefully provides a solution for the problem. The chart is then studied carefully to see if the logic of the steps in the program will allow for every situation created by the data. When the chart is thought to be perfect, the next step is taken.

Step 2. The programmer uses the chart to construct a program in one of the translation languages such as FORTRAN, COBOL, BASIC, or PL/1.

Step 3. The translation language program is punched on cards or typed on a teletypewriter and compiled by the computer. That is, the computer translates the translation language into machine language. If punched cards are used, the machine language deck will be quite thick compared to the translation language deck. Every translation language command represents many machine language instructions.

66   Introductory College Mathematics

Step 4.   The data and the program are put into the computer
and the program is run. Assuming that the program
is bug free, the program will be completed.

Step 5.   The answers required by the program are decoded
from the 1's and 0's used in the computer and
printed by the OUTPUT printer or recorded on
tape as dictated by the program.

Exercise Set 4

1. For each flowchart write a program in BASIC language.

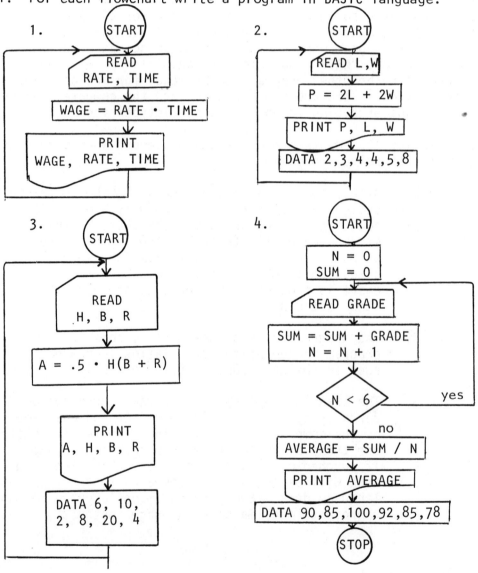

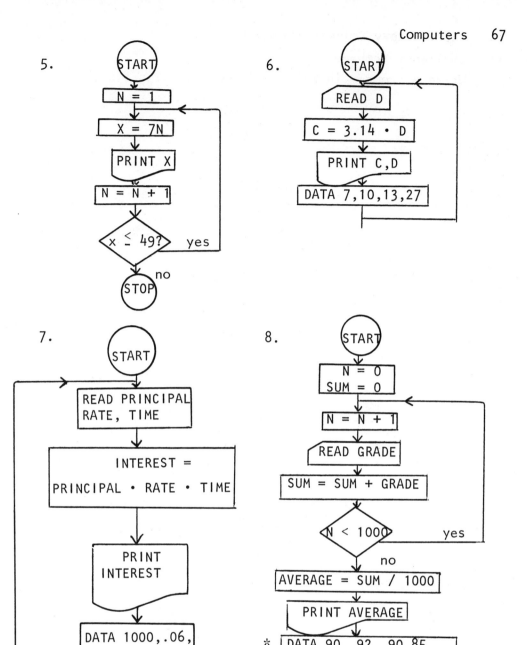

68   Introductory College Mathematics

11. **Challenge Problem**

Write a program in BASIC for the following flowchart:

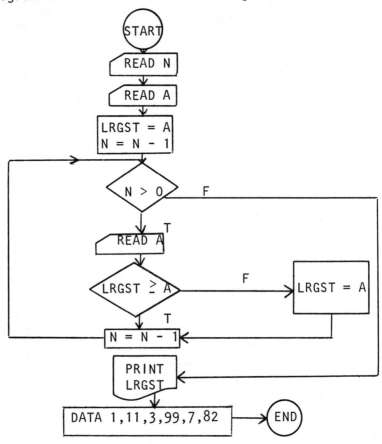

MODULE SELF-TEST

1. A Babbage "Difference" array is shown for generating perfect squares. Construct a "Difference" array for perfect cubes.

Number	Square	Difference	
1	1		
		3	
2	4		2
		5	
3	9		2
		7	
4	16		2
		9	
5	25		2

2. Which of the following is an "and" gate?

   a.

   b.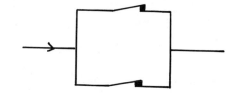

3. Will current flow through the circuit below?

   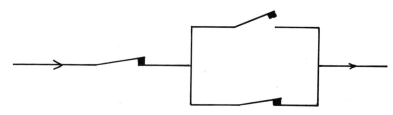

4. Add the following base two numerals:

   a. 1101_2 b. 1001_2
        $\underline{101_2}$ $\underline{111_2}$

70   Introductory College Mathematics

5. If 1 and 0 are inserted at lines a and b, respectively, on the half-adder, find the states at: c, e, f, i, j, and k.

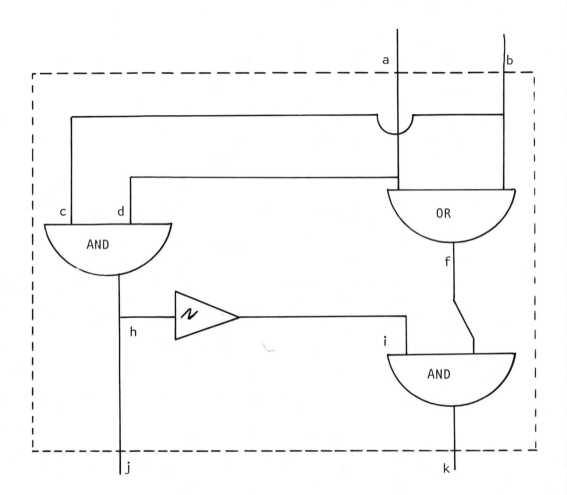

Computers 71

6. If 1, 0, and 1 are inserted at points a, b, and c respectively on the full-adder, find the states at: d, e, f, and g.

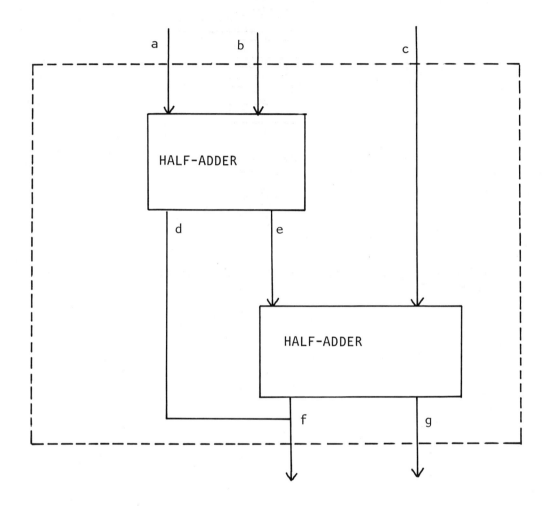

72   Introductory College Mathematics

7. What is the process shown by the following flowchart?

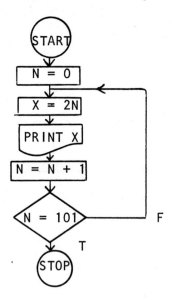

8. Write a program using BASIC language for the flowchart in problem 7.

PROGRESS TEST ANSWERS

Progress Test 1

1. Table of Cubes

Number	Cube	Difference
1	1	
		7
2	8	12
		19	6
3	27	18
		37	6
4	64	24
		61	6

Progress Test 2

1. True
2. False
3. False
4. True
5. False
6. True

Progress Test 3

1. 11_2
2. 10_2
3. 1_2
4. 11_2
5. 10_2
6. 1000_2

Progress Test 4

1. 0, 1, 0, 1
2. 0, 1, 0, 1
3. 0, 0, 0, 0

Progress Test 5

1. 11, 1, 0
2. 10, 1, 0
3. 10, 0, 1
4. 01, 0, 1

Progress Test 6

<u>Filing and Information Retrieval</u>
Personnel files
Computerized Criminal History Program
Airplane reservations

<u>Data Manipulation</u>
Weather plotting
Translating languages
Typesetting

<u>Simulation</u>
Studying a design before an object is built
Studying a cultural situation and recommending changes

<u>Control</u>
Construction cost control
Machine Control
Automatic washing machine

<u>Pattern Recognition</u>
Sorting

<u>Calculations</u>
Scientific computations

<u>Heuristic Problem Solving</u>
Analysis of political systems
Games

74   Introductory College Mathematics

Progress Test 7

1. His brain.
2. a. a, b, d, e
   b. a, b, d
   c. b, d, e
   d. a, b, d
   e. a, b, d, e

Progress Test 8

1. a.
   b.
   c.

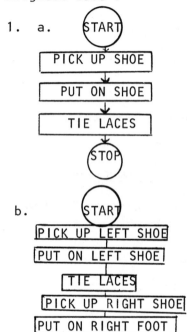
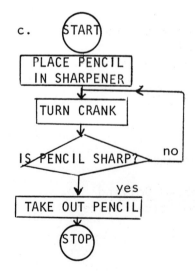

Progress Test 9

1. Telephoning for a date.
2. Printing the whole numbers from 0 to 999 inclusive.
3. Printing the whole numbers 0,1,2,3,...
4. Finding the solution for $y = 3x + 1$ when $x = 0$

Progress Test 10

1. A = L * 0.01
2. 0
3. SUM = SUM + A
4. N = 0 and N = N + 1
5.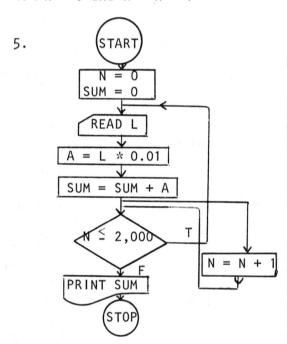

Progress Test 11

1.
```
10 REM FIND CIRCUMFERENCE
20 READ D
30 LET C = 3.14 * D
40 PRINT C
50 GØ TØ 20
60 DATA 7,8,9
70 END
```

Computers   75

Progress Test 11 (continued)

2.  ```
    10 REM FIND DISTANCE
    20 READ R,T
    30 LET D = R * T
    40 PRINT D
    50 GØ TØ 20
    60 DATA 60,1,30,2,90,3
    70 END
    ```

3.  ```
    10 REM SUM NUMBERS FRØM
    20 REM 0 TØ 100
    30 LET X = 0
    40 LET S = 0
    50 LET S = S + X
    60 LET X = X + 1
    70 IF X < 100 THEN 50
    80 PRINT S
    90 END
    ```

EXERCISE SET ANSWERS

Exercise Set 1

I. 1. T   2. F
    3. F   4. T
    5. T   6. T
    7. F   8. F
    9. F   10. T
    11. F 12. T
    13. F 14. T

    15. a.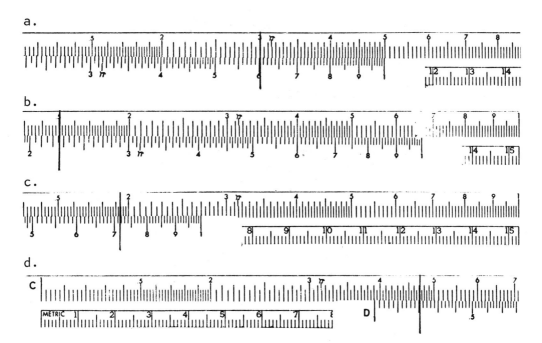

    16. 164.8 seconds   17. 1.648 seconds

II. *Challenge Problems*

a.
b.
c.
d.

76 Introductory College Mathematics

Exercise Set 1 (continued)

e.
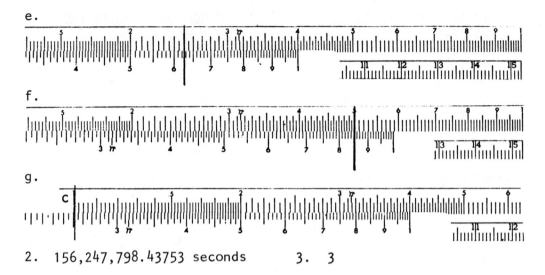

f.

g.

2. 156,247,798.43753 seconds   3. 3

Exercise Set 2

I. 1. True
    2. a. no   b. no   c. no   d. yes   e. yes   f. no

    3. a.  yes

        b.  yes

        c.  yes

        d.  yes

        e.  no

Exercise Set 2 (continued)

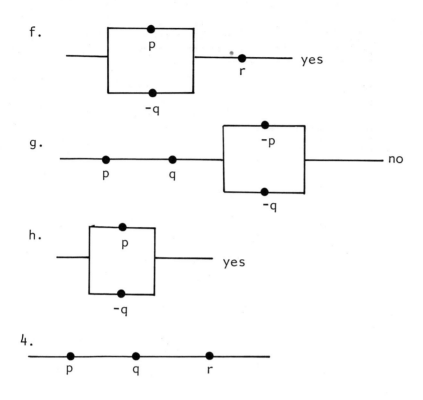

4.

5. a. 10_2 b. 11_2 c. 101_2 d. 1100_2 e. 10_2 f. 1111_2 g. 10000_2
    h. 100_2

6. fg,   d,   e
    a. 10,   0,   1
    b. 01,   0,   1
    c. 01,   0,   0
    d. 00,   0,   0

II. *Challenge Problems*

1.

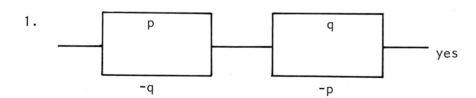

78   Introductory College Mathematics

Exercise Set 2 (continued)

2.

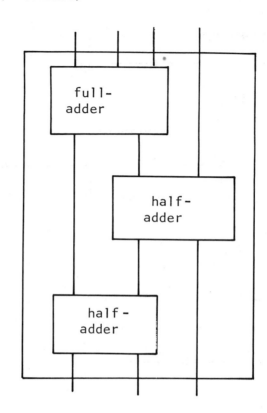

3.

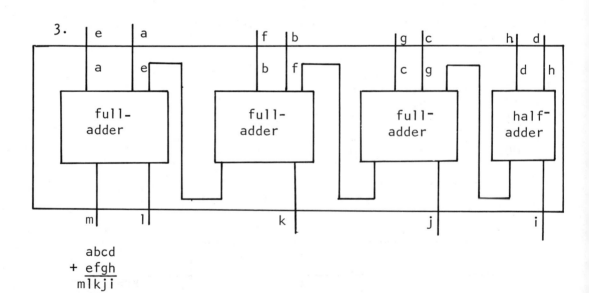

```
  abcd
+ efgh
  mlkji
```

Exercise Set 3

1.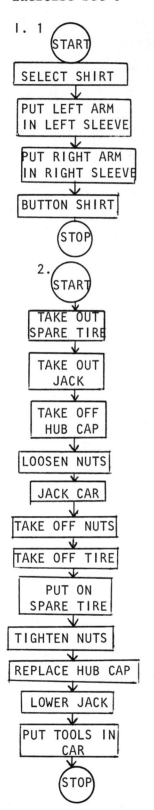

2.

3.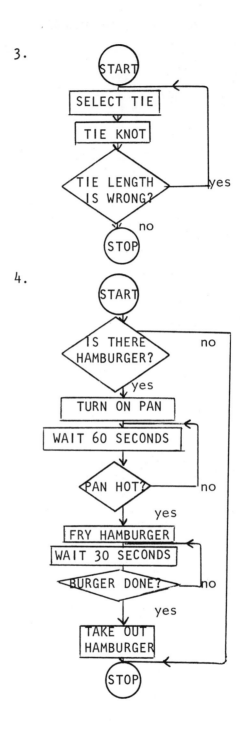

4.

80   Introductory College Mathematics

Exercise Set 3 (continued)

5.

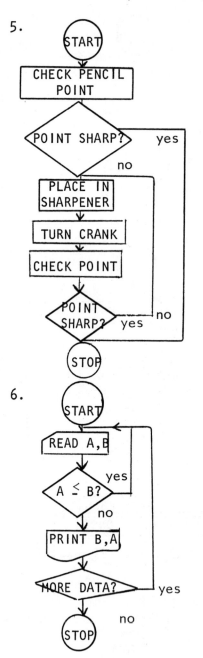

6.

7.

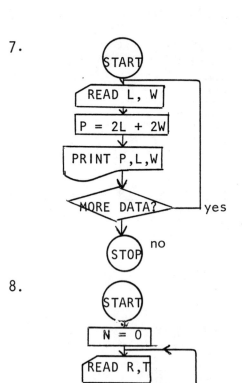

8.

9. Printing the first 1,000 prime numbers starting with 2.

10. Printing the even numbers from 2 to 100 inclusive.

11. Printing the area, length, and width of some rectangles.

Exercise Set 3 (continued)

12. Printing the numbers 3, 9, 27.
13. Printing the multiples of 6 from 6 to 66 inclusive.
14. Printing the pairs (0,5), (1,7), (2,9), (3,11), (4,13), (5,15), (6,17), (7,19), (8,21), (9,23), (10,25)
15. Printing the sum of the areas of some rectangles with width 0.01.

Exercise Set 4

I. 1.
```
10 REM FINDING WAGES FROM
20 REM RATE, TIME
30 READ R,T
40 LET W = R * T
50 PRINT W,R,T
60 GØ TØ 30
70 END
```
2.
```
10 REM FIND PERIMETER
20 READ L,W
30 LET P = 2 * L + 2 * W
40 PRINT P,L,W
50 GØ TØ 20
60 DATA 2,3,4,4,5,8
70 END
```
3.
```
10 REM AREA
20 READ H,B,R
30 LET A = .5 *H * (B+R)
40 PRINT A,H,B,R
50 GØ TØ 20
60 DATA 6,10,2,8,20,4
70 END
```
4.
```
10 REM GRADE AVERAGE
20 LET N = 0
30 LET S = 0
40 READ G
50 LET S = S + G
60 LET N = N + 1
70 IF N < 6 GØ TØ 40
80 LET A = S / N
90 PRINT A
100 DATA 90,85,100,92,85,78
110 END
```

5.
```
10 REM MULTIPLES ØF 7
20 LET N = 1
30 LET X = 7 * N
40 PRINT X
50 LET N = N + 1
60 IF X ≤ 49 THEN 30
70 END
```
6.
```
10 REM CIRCUMFERENCE
20 READ D
30 LET C = 3.14 * D
40 PRINT C,D
50 GØ TØ 20
60 DATA 7,10,13,27
70 END
```
7.
```
10 REM FIND INTEREST
20 READ P,R,T
30 LET I = P * R * T
40 PRINT I
50 GØ TØ 20
60 DATA 1000,.06,2,2000,.12,1
70 END
```
8.
```
10 REM AVERAGE GRADES
20 LET N = 0
30 LET S = 0
40 LET N = N + 1
50 READ G
60 LET S = S + G
70 IF N < 1000 THEN 40
80 LET A = S / 1000
90 PRINT A
100 DATA 90,92,93,85,60,...
110 END
```

11. *Challenge Problem*

```
10 READ N
20 READ A
30 LET L = A
40 LET N = N - 1
50 IF N ≤ 0 THEN 110
60 READ A
70 IF L ≥ A GØ TØ 90
80 LET L = A
90 LET N = N - 1
100 GØ TØ 50
110 PRINT L
120 DATA 1,11,3,99,7,82
130 END
```

MODULE SELF-TEST ANSWERS

1.
Number	Cube	Differences
1	1	
		7
2	8	12
		19	6
3	27	18
		37	6
4	64	24
		61	6

2. a   3. yes   4. a. 10010_2 b. 10000_2

5. $c = 0$, $e = 0$, $f = 1$, $i = 1$, $j = 0$, $k = 1$

6. $d = 0$, $e = 1$, $f = 1$, $g = 0$

7. Printing multiples of 2 from 0 to 200 inclusive.

8.
```
10 REM MULTIPLES ØF 2 FRØM 0 TØ 200
20 LET N = 0
30 LET X = 2 * N
40 PRINT X
50 LET N = N + 1
60 IF N < 101 THEN 30
70 END
```